CAREERS IN
INDUSTRIAL DESIGN

PRODUCT DESIGNER

LOOK AROUND AND WHAT DO YOU SEE? A smart phone, a chair, a lamp, a kitchen appliance, and maybe a vacuum cleaner. What do these items have in common? They are all examples of industrial design that started as ideas drawn on a sketchpad. They look and function as they do because an industrial designer created them that way. You may not recognize the names Jonathan Ive, Earl Dean, or Egmont Arens. You would instantly recognize an iPod, Coke bottle, or KitchenAid mixer, which they designed – items that have been elevated to iconic status in American culture.

Industrial designers develop and design manufactured products, such as sporting goods, appliances, toys, cars, and consumer electronics. Nearly everything we see and touch in everyday life has been conceived by an industrial designer. That is an uncountable number of products! So many, in fact, that most industrial designers specialize in one particular product category, such as office chairs, running shoes, or kitchen faucets.

The goal for every industrial designer is to seamlessly blend form and function to make a product desirable in every way – looks, usability, and cost and ease of manufacturing. It is a crucial balance that sets industrial design apart from other types of design. The products industrial designers create must be attractive to customers, but it is equally important that they be usable, comfortable, high quality, affordable, and safe. This requires a combination of artistic skills and technical knowledge of materials, ergonomics, costs, and manufacturing processes.

The process of industrial design is multifaceted and rarely accomplished by just one person. Industrial designers usually work in teams with other designers, as well as professionals from other disciplines. A typical team might include engineers, materials scientists, cost estimators, usability experts, marketing specialists, sales staff, brand developers, and market researchers. The process starts with the industrial designer, whose job is to come up with ideas for new products or ways to improve existing products. The designer presents concepts through sketches or computer renderings, models, and prototypes. The other team members help figure out what materials to use, how much manufacturing will cost, and how the product will be marketed.

A good imagination and artistic talent are prerequisites to working in industrial design, but a college education is needed to be fully prepared for this career. Most employers prefer applicants who have a bachelor's degree in industrial design. However, an engineering degree is also acceptable so long as basic art and design courses have been included in the curriculum. Some graduates choose to continue their education and obtain higher degrees that will make them more attractive to employers. Earning a master's degree in business administration (MBA), for example, is the best way to learn about marketing, quality control, accounting, project management, and strategic planning. It also helps a designer qualify for management positions.

Job opportunities can be found in every industry. Since new products and innovations are introduced every day, and in almost every category of consumer goods, from baby bottles to refrigerators, good industrial designers are always in demand. The niche in which they find success often depends on related personal interests and related job skills, but choosing a hot specialty can make all the difference when setting out on this career path. Currently, industrial designers working in consumer electronics, transportation, and especially medical equipment, are experiencing the highest demand.

Industrial design is a great choice for the creative person who can figure out how things work. Do you use both sides of your brain equally? If so, this profession could be a great fit. The work is challenging, but it's also stimulating and fun, and the pay is good. If you think it would exciting to see your ideas become real products used by millions of people, read on to learn more about careers in industrial design.

WHAT YOU CAN DO NOW

THIS CAREER REQUIRES A COLLEGE EDUCATION. Plan accordingly when designing your high school curriculum. To make sure you fulfill all the admissions requirements, contact the colleges of your choice, and check with your guidance counselor. Take Advanced Placement (AP) courses whenever possible, especially in math and science. Chemistry and physics, in particular, are becoming increasingly important in this field. AP courses will strengthen your college application, plus you might receive college credit for them.

Beyond required courses, the focus should be on art and computer classes. Drawing, painting, and sculpture would be a great start. Shop classes are another good choice. Working hands-on with metal and wood will teach you how to use machines and tools to manipulate materials. If you are lucky enough to be in a high school that offers a graphic design course, take it. If not, there are many tutorials on the Internet, and you can teach yourself any of the popular graphic design software programs. Industrial designers routinely use Computer-Assisted Design (CAD) programs to create graphics and simulate 3D objects, so the sooner you get training in CAD the better. If your high school has a CAD class, great. If not, look for classes online or extension courses through your local community college.

Giving presentations to clients and colleagues is an important part of an industrial designer's job. A class in public speaking or joining the debate club will help you gain the necessary skills and confidence. If you think you might want to run your own design firm someday, consider taking business classes such as accounting and marketing.

Pursue any craft-related hobbies that will help you develop artistic skills, such as jewelry making, woodworking, sculpting, or pottery.

Ask your guidance counselor to help arrange a visit to a factory that has an industrial design department. Your counselor might also be able to arrange for you to job shadow an industrial designer for a day. That would be a great way to discover what the day-to-day work is really like.

Read industry publications, such as Design News, to keep up with new techniques and current trends.

HISTORY OF THE CAREER

IT IS WIDELY ACCEPTED THAT INDUSTRIAL design was a product of the Industrial Revolution. For several millennia the creation and production of products were the responsibility of individual craftsmen. During the Medieval period, trade grew beyond local towns, opening up opportunities for specialist craftsmen to come together in groups that would share training and techniques. Large workshops emerged in cities like Florence, Venice, and Nuremberg. By the early 16th century, there were numerous pattern books and collections of engravings that illustrated how common objects could be produced through the repetitive duplication of models.

In the 17th century, many of these groups were installed in large government-operated manufacturing operations in France and other countries. In factories like the famous Gobelins Manufactory in Paris, teams of hundreds of craftsmen created products ranging from furniture to tapestries to coaches. This pattern of large-scale production continued into the early 18th century with porcelain factories such as the Meissen workshops in Germany and Royal Worcester in England.

In the US during the late 19th century, products and innovations were mostly concerned with functionalism and modern industrial mass production. Due to increasing labor costs and a shortage of cheap labor, US manufacturers focused on standardization – the hallmark of the Industrial Revolution. By the end of the century, industrial design had become incorporated into industry. However, the profession was undefined and any worker could be labeled as an industrial designer. This often included artists, architects, craftsmen, inventors, engineers, technicians, and even some business types such as marketing and salespeople. It was not until the 1920s that the field of industrial design developed fully and integrated all of these activities into one distinct profession.

By the early years of the 20th century, the majority of everyday products were routinely mass-produced. Individual manual labor was replaced by assembly lines, and eventually robotics. The crafts-based economy of the past had ended. Artists and craftsmen were left out of the manufacturing process, replaced by designers who would figure out what and how objects could be mass-produced.

American companies were slow to catch on to the importance of style and design. In the US, the goal throughout the Industrial Revolution was to produce functional products in the most efficient and least expensive way possible. But the average American shopper wanted more. Consumers were drawn to products that were affordable, yet aesthetically pleasing. They welcomed styles that changed with the seasons. Boring design led to sagging sales and sometimes failure in the marketplace. The most successful manufacturer of the Industrial Revolution, Henry Ford, learned this lesson the hard way. For nearly 20 years, Ford built only one car, the Model T. The car was functional and easy to produce on Ford's assembly lines. Fifteen million of them were churned out cheaply, but they were anything but attractive. Some even called them ugly, saying they looked like tin cans on wheels. Ford, however, paid no attention to the criticisms. After all, he was selling more cars than anyone else. To show his disdain for attractive design, he famously said, "Any customer can have a car painted any color that he wants so long as it is black."

In 1926, General Motors introduced its stylish new Chevrolet touring car. It was a smash hit with car buyers and for the first time, Ford was outsold by a competitor. Only then, did he realize that good design was good business. More and more companies caught on to the importance of style and design, and by 1927, manufacturers were routinely employing industrial designers to make products and packaging more stylish and attractive to consumers. Throughout the latter half of the 20th century, the field of industrial design grew quickly.

Since the 1950s, industrial designs have made a significant impact on American culture and daily life. A number of industrial designers have contributed iconic designs that have been noted and documented by historians of social science. Here are just a few of them:

Viktor Schreckengost was an American artist, sculptor, and teacher turned industrial designer. His wide-ranging work included everything from pottery designs to lawn chairs and dinnerware. As an artist, he created the largest freestanding ceramic sculpture in the world, titled Early Settler. As an industrial designer, he is best known for the banana seat bicycles manufactured by Murray bicycles for Murray and Sears, Roebuck and Company. He also designed the first truck with a cab-over-engine configuration, a design in use to this day. Schreckengost was the founder of The Cleveland Institute of Art's school of industrial design.

Richard Teague spent most of his working life designing cars for the American Motor Company. His original concept using interchangeable body panels allowed him to create a wide array of different vehicles using the same stampings. That was a win-win for AMC and style-conscious car buyers alike. Some of his most unique automotive designs included the Pacer, Gremlin, Matador coupe, Jeep Cherokee, and Eagle Premier.

Perhaps the most influential industrial designer of our time is **Sir Jonathan Ive.** As the Chief Design Officer of Apple, he oversees the Apple Industrial Design Group and provides leadership for the Human Interface software teams company wide. Ive is personally responsible for the design of many of Apple's iconic products, including the MacBook (Air and Pro), iMac, iPod, iPhone, iPad, and Apple Watch.

Today, industrial design continues to grow as a profession as the scope of its influence expands to include everything from designing the tools to produce a product to designing its shipping container that will get it into the hands of eager consumers.

WHERE YOU WILL WORK

THERE ARE CURRENTLY MORE THAN 40,000 industrial designers working in the United States. The highest concentration of industrial designers are located in Indiana, California, Michigan, Pennsylvania, Rhode Island, and Wisconsin. The two states that employ the most – more than 3,000 industrial designers each – are California and Michigan.

Nearly one third of all industrial designers work in manufacturing, developing concepts for products we use every day. About 30 percent of them are employed directly by manufacturers, while a smaller number are self-employed designers working under contract. Aside from manufacturing, industrial designers work mostly in specialized design services firms, corporate research and development departments, and engineering firms. About one in four are self-employed.

Most industrial designers specialize in a particular type of product. There are many choices, but some of the most common include:

- Toys and games
- Household appliances
- Consumer electronics

- Furniture
- Sporting goods
- Machinery, tools, and manufacturing equipment
- Biomedical equipment
- Motor vehicles
- Clothing and shoes
- Phones

Industrial designers work in offices with comfortable surroundings, good lighting and ergonomic furniture. A typical workspace includes drafting tables for sketching designs, computers and other office equipment for preparing designs and presentations for clients, a meeting room with whiteboards for brainstorming sessions with associates, and a conference room for meeting with clients. Self-employed designers work in their own offices, which are often located in their home.

Travel is common in this career. Industrial designers typically meet with clients in their homes or offices, showrooms, production facilities, and trade shows or other exhibition sites. They may also travel to locations for product testing, such as design centers, dedicated testing facilities, offices of clients, supplier warehouses, or places where products are manufactured. Most of this travel is domestic, but corporate industrial designers travel the world to meet with clients, source supplies, and oversee certain manufacturing processes for quality control.

Work Schedules

Industrial designers usually work full time. The hours are steady for those who are employed by manufacturers, large corporations, or design firms where the usual five-day, 40-hour workweek is the norm. However, designers who work for smaller companies may occasionally need to work longer hours to meet deadlines.

Consultants and designers who are self-employed may need to frequently adjust their schedules to meet with clients in the evenings or on weekends. In addition, they may spend a significant portion of their time looking for new projects or bidding against other designers for contracts.

THE WORK YOU WILL DO

AN INDUSTRIAL DESIGNER IS THE creative force behind all manufactured products, from toothbrushes to snowboards, dishes to motor scooters. They provide the vision that can take the idea of an inventor and turn it into a marketable product that is consumer friendly and fills a need. The need may be real, but it is just as likely to simply reflect the style and trends of the moment. In either case, the best designs are a perfect balance of form and function.

It takes a combination of artistic, business, and engineering skills to create products that people use every day. Superior artistry can capture the attention of a consumer while perpetuating a particular brand's attitude and style. Business savvy is used to analyze production costs and market placement while maintaining a company's goals and culture. Engineering expertise is needed throughout the production process, from assessing the feasibility of making a product to determining the most efficient equipment for manufacturing.

Industrial designers may work on creating new products or improving existing ones. Most of the time, industrial designers develop concepts for new products that are manufactured in factories. This includes a wide range of items, from everyday products like toys, appliances, and furniture, to larger things like industrial equipment, motor vehicles, and heavy machinery. They are also trained to utilize emerging technologies, materials, and manufacturing methods to improve the design and usability of items already in the marketplace. They may revamp an item to enhance its features or increase consumer appeal by updating the style.

Industrial designers tend to work within a particular industry, creating certain types of products, such as sporting goods, housewares, or medical devices. Some do not focus on the products themselves. Instead, they are branding experts who create packaging, exhibits, or shipping containers. Industrial designers are behind the familiar look of iconic products like iPods, Jeep, and the Coke bottle.

No matter what type of work an industrial designer chooses, the process always begins as a concept in someone's mind. Making that concept a reality is a long and involved process. Actual tasks may vary, but the work of an industrial designer typically involves doing the following:

- Research how the product might be used and who will want to buy it

- Sketch out ideas

- Use computer software to develop virtual and/or 3D models of proposed designs

- Select materials and determine production costs for manufacturing

- Consult with other specialists to evaluate technical concepts, such as ergonomics, sustainability, or market placement

- Evaluate product aesthetics and function to determine if a design is practical

- Perform safety tests

- Present designs and demonstrate prototypes to clients for approval

For every proposed idea, an industrial designer will come up with a multitude of designs. Only those products that can be easily manufactured and that are cost effective are likely to be mass-produced.

The process of design is considered creative, yet in the field of industrial design there are many analytical steps that must take place. Some of the more commonly used are comparative product research, prototyping, user experience evaluation, and testing. Much of the hands-on work is accomplished with the aid of computers. Designs may first be sketched by hand, but industrial designers are more likely to use a computer-aided design (CAD) program to finalize an idea. The product idea is then reviewed by brand managers and top-level executives before being accepted and mass-produced.

Tools of the Trade

Computers are an important tool in industrial design. Two-dimensional computer-aided design (CAD) software is routinely used to sketch ideas. Hand- sketching is still done, but CAD software is much faster. It also makes it easier to make changes and show alternatives. Three-dimensional software can transform digital sketches into physical models with the help of 3D printers.

Industrial CT (computed tomography) scanning has an important role in the manufacturing process. It can be used to test prototypes both internally and externally, but its key function is to perform internal inspections of components to detect flaws or analyze potential failures. Once a product prototype has been tested for defects, the manufacturing process can be modified to improve the product, the tools used to make it, and the assembly procedures.

Industrial designers with engineering experience may use computer-aided industrial design (CAID) software to create instructions that manufacturing equipment can read and follow to build a product to precise specifications.

Teamwork

Industrial designers rarely work alone. The work usually requires a high level of collaboration with other designers and people from other disciplines. They typically work in teams that may include specialists like materials scientists, engineers, feasibility advisors, tooling and production experts, and testers. They may work with people in other departments at the design firm as well as people from the client company. Client groups are usually comprised of management, marketing specialists, sales staff, accountants, and cost estimators. With all this input, industrial designers can determine if a proposed product will be safe, functional, attractive to customers, and affordable to make.

Specialization

Industrial designers usually focus on one product category, such as sporting goods, toys, housewares, or medical equipment. The choice of specialty is often dictated by the designer's own interests. For example, a designer who loves winter sports might want to design snowboards. An amateur chef might develop ideas for stylish cookware or more versatile kitchen appliances. Because there are so many different kinds of products, there is often overlap that creates subspecialties. An industrial designer working on consumer electronics, for instance, might specialize in designing smartphones. A furniture designer might concentrate on ergonomic office furnishings. Designers employed by manufacturers are limited to designing products that fit into their company's line. Self-employed designers have more flexibility in the product categories they work on.

The Process

Long before picking up a sketchpad, industrial designers must ascertain the purpose of the new product. The very first step in creating a new design involves research. This always includes meeting with the client to discuss the requirements and goals of the business. There may be numerous strategy sessions with the corporate staff who will stipulate guidelines such as budgets and deadlines. To learn more about prospective customer tastes, the designer may consult with the marketing staff, conduct independent market research, read consumer reports, attend trade shows, analyze competitor products, and visit with suppliers and manufacturers of related products.

Next comes the concept design phase. Decisions are made regarding specific characteristics such as size, shape, weight, and color. At this time, product specifications are still fluid. The design team usually comes up with several variations, which will be explored in the next step.

It is now time for the designers to illustrate a vision for the product. Sketches may be created by hand drawing or by preparing digital sketches with a computer-aided design tool (CAD). In addition to putting ideas to paper or screen, hand-built models may be produced using cardboard, plastic, clay, or foam.

Promising sketches are collected and organized into one place for review by the senior designer and design director. The purpose of this internal review is to begin selecting the best ideas. Once the ideas have been filtered and refined, each design sheet is scaled to be approximately the same size so they will all have the same visual impact when presented to the client.

Next, the sketches are shown to the client. After the client has decided what ideas are to be pursued, the design team will create 3-dimensional mock-ups to help clients visualize the final results. The prototypes are then presented to the client, who may make suggestions or ask for changes.

After the client decides on the final product design, the project moves into the testing and refinement stage. At this time, a number of prototypes will be built and evaluated for appearance, function, and safety. Sample prototypes are given to consumers or beta testers who will report on any problems that need to be addressed prior to the product's completion or manufacture. Safety is an important factor to consider. A toy for a toddler, for example, cannot have parts that might be swallowed. Nor can a lamp get so hot it could catch on fire.

STORIES OF INDUSTRIAL DESIGNERS

I Design High Quality Kitchenware

"My designs are a balance of aesthetics and ergonomics. My teammates and I start each design project by sketching rough ideas onto paper. When we think we've got some keepers, we narrow them down to a handful of the best. We meet with the client, and the client chooses their favorites. We then take those and develop 3-D mockups for a second presentation. Once the client makes a final choice, we refine the design until it's ready to move into engineering.

I originally studied mechanical engineering because I like to create tangible objects. But I've always like drawing. It wasn't until I went to a design show that I realized I could do both by becoming an industrial designer. Making the switch was the best decision I could have made.

The artist in me has the most fun during the first phase of the project, which is concept sketching. I let my mind run free and draw things I think look good or will appeal to consumers in the market we're trying to hit. It is essentially doodling at this point. Sometimes I can't believe I get paid for it. It doesn't really feel like work. The next phase, which is meeting with clients, is more challenging for me. Every presentation has to include specifics about each individual concept. I have to describe the technical features and explain why I made it look a certain way. It can be hard to get clients to understand my design intent. I didn't pay enough attention to communication in school so now it is something I continually have to work on. Great communications skills are crucial as a designer and are harder to develop when you're learning on your own.

My advice to aspiring industrial designers is to get as much experience in the field as possible, as early as possible. My school offered a work co-op program, but you could get internships that will be just as useful. With the advice of my counselor, I worked in five different companies before I graduated. The program was a huge asset. Not only did I learn valuable lessons that were not presented in the classroom, but I got my first job at one of those companies."

I Am a Digital Product Designer

"I design mobile apps. In this field, good work is invisible. You know it's a good design when you touch a button and the app takes you where you need to be without any thought on your part. For my part, a ton of thought goes into making that happen. In fact, the majority of my time is spent thinking about the people who will use the products I'm designing. I work really hard to understand the end users, what they want most, and how to avoid any frustrations from using the product. This is a job for thinkers. A lot goes into designing products, whether they are digital or hard products. I work closely with members of my team, from creating strategy to developing user flows to the actual building of the product.

Many talented people who would do well in this career are missing out because they don't understand what it's about. They hear the word 'designer' and think it means making things pretty – something that might turn off an engineering student, for example. The word 'industrial' also confuses people. It sounds like I work in a factory, making machinery or equipment. As a term, 'industrial design' seems too general and maybe a little vague. That's why a lot of industrial designers prefer to use the term 'product designer' when talking to people outside of the field.

The best designers think about design 24/7. That might sound excessive or obsessive, but it's a good habit to get into. I am constantly studying user research techniques, interaction patterns, and new prototyping tools. And I make a point to expose myself to great design every day. Reading about ingenious examples of industrial design that combine aesthetic appeal with creative problem solving helps me become a better designer. Anyone who is serious about this career should take advantage of the many resources that are out there.

Aspiring industrial designers should practice and practice a lot. Look for problems that you find frustrating and see what you can do to solve them. Think an app on your Android phone could be better? Design a new one. The tools are out there for anyone to use. Prototype your new version and ask your friends for feedback. There are also some great online design communities like dribbble.com, where you can go to swap ideas with other designers."

PERSONAL QUALIFICATIONS

GOOD INDUSTRIAL DESIGN IS A balanced blend of art and science, making it a great fit for anyone who can use both sides of the brain equally. The technical knowledge, which is primarily being able to use computer-aided design software to develop designs and prototypes, can be obtained through education and training. The necessary artistic ability is usually innate, but can also be acquired through instruction and practice.

Artistic ability starts with creativity and a good imagination. Creativity in the world of product development is about being innovative and also finding ways to seamlessly integrate technologies into new products. Basic drawing skills are needed to express creative ideas. Although CAD software is used throughout the design process, industrial designers sketch their initial designs first. Sculpting skills are then needed to produce 3-dimensional models or prototypes. Good designers have a keen sense of style, and an eye for color, proportion, balance, and detail.

Unlike most artistic fields, industrial design requires some mechanical aptitude. It is important to understand how manufactured goods are engineered, from concept to finished product. Products must be stylish, of course, but they must also perform a purpose in the real world. The best designs are a perfect balance of form and function. Successful industrial designers have a good understanding of how things work, including the technical aspects involved in manufacturing the products they design.

Analytical skills are essential. Long before picking up a sketchpad, the industrial designer must study consumers to determine their habits, tastes, and needs. The decision to bring a new product to market is very difficult an requires intellectual depth. Good problem-solving skills can help pinpoint complex details such as optimum size, manufacturing cost per unit, alternative materials, retooling needs, and other production issues.

Successful industrial designers are self-disciplined. Although most work in collaboration with others, every designer has to be able to work independently with minimum supervision. Time management is key, from getting projects started without delay to keeping up with production schedules to meeting final deadlines. The ability to work under pressure is an important trait. If a designer becomes overwhelmed or loses focus, the quality of work can suffer.

Good communications skills, both visual and verbal, are important. Communication between industrial designers and clients is done through presentations. Those who are able to clearly communicate their ideas verbally and visually are the most comfortable with making presentations, and those who make good presentations are the most successful in their careers. Good people skills are important, too. In addition to developing relationships with clients, industrial designers must have good working relationships with other designers and engineers who may be working on the same project.

It is common for industrial designers to pursue self-employment at some point. Going freelance or starting a firm requires a good head for business, some sales ability, and a knack for convincing clients to accept new ideas and back projects. Knowledge of accounting, marketing, purchasing, project management, and quality control is useful. Training in these areas can easily be obtained through online or community college courses.

ATTRACTIVE FEATURES

ACCORDING TO THE NATIONAL Endowment for the Arts, there has never been a better time to be an American industrial designer. American companies have fully accepted the importance of design, and the investment they are willing to make is being felt in every industry. For designers, this translates to getting paid well to be creative and innovative.

Industrial designers earn good money. The average salaries for staff designers start out at a respectable $40,000 for new graduates and continually grow from there. Many experienced industrial designers earn six figure incomes, not counting benefits. Some even receive bonuses and profit sharing on top of their base salaries. Self-employed designers, including freelancers and those who own their own design firms can earn even more than their salaried colleagues. However, they do have to provide their own benefits, and keeping the work flowing in can be a full-time job in itself.

This is a career that offers considerable flexibility. You can take a secure job as a corporate employee, or you can retain your independence and freelance. Corporate employees enjoy regular hours and steady paychecks. Freelancers can work when and where they choose, and the ambitious types can potentially out-earn anyone in a staff position.

There is also flexibility in the type of work. Industrial design crosses every industry, creating a wide variety of jobs. Industrial designers usually focus on a particular type of product. For example, an avid runner might specialize in the design of athletic shoes. Or someone interested in improving healthcare could specialize in designing better medical devices. There are countless choices from packaging to furniture to the tools and machinery needed to make products. Because the skill set is essentially the same for all industrial design, it is possible to change fields during your career.

Best of all, industrial designers get to be creative. The whole point of their job is to come up with new ideas and turn them into tangible products that people want. For the artistically inclined, what could be better than drawing pictures and getting paid for it? Some designers say the work is so much fun, it doesn't feel like work at all.

UNATTRACTIVE ASPECTS

INDUSTRIAL DESIGNERS ARE GENERALLY content with their career choice. Most report a high level of job satisfaction, but no profession is perfect. The most common problem is deadlines. Deadline demands create pressure and longer working hours. This happens most often to self-employed designers who have to juggle several clients at once. It also happens when designs are rejected, or prototypes do not turn out as planned, or simply because the creative juices refused to flow on demand. It is nice to get paid for being creative, but having to be creative on demand can be stressful.

Freelancers tend to have more challenges than staff designers. They cannot count on a regular work schedule or even a regular paycheck. They have to figure out how to keep money coming in because they are paid by the project. That often means rushing to finish a job quickly while at the same time, preparing proposals to get new clients.

Designers cannot expect everything they design to be produced. Ideas are just ideas, and not all of them are going to fly. Designs have to be based on the client's preferences, which may be quite different than your own ideas. As an industrial designer, you will face critics every day. Rejection is part of the job and it cannot be taken personally. Your new idea might be great, but perhaps the cost of material and manufacturing is too high. Many clients do not like change. They prefer to stick with the familiar and safe, rather than be on the cutting edge with something daring and innovative.

EDUCATION AND TRAINING

A BACHELOR'S DEGREE IS USUALLY required for entry-level industrial design jobs. Most employers prefer a bachelor's degree in industrial design, architecture, or engineering. A bachelor's degree program in industrial design is typically offered through architecture, design, or art colleges, usually within larger universities.

The National Association of Schools of Art and Design (NASAD) accredits about 320 colleges, universities, and independent institutions with programs in art and design. Of these, 45 offer degree programs in industrial design.

It can take up to five years to earn a Bachelor of Industrial Design (BID) degree. A Bachelor of Fine Arts (BFA) or Bachelor of Science (BSc) degree can be obtained in four years, but depending on the type of industrial design work, the degree may not be enough to qualify a candidate for a job.

Getting accepted into an industrial design program is not as easy as getting into a traditional art school. Most schools require successful completion of some basic art and design courses. You may also need to submit sketches and other examples of your artistic ability.

Industrial design programs vary, but the curriculum is always comprised of classroom instruction and hands-on training in computer labs and design studios. Students are also encouraged to create independent design projects, and pursue practical training experiences.

Most programs include courses in conceptual sketching, computer-aided design and drafting (CADD), industrial materials and processes, 3-dimensional modeling, and manufacturing methods. Other typical courses cover topics such as business, visualization, ergonomics, sustainable design, interactive design, creating models and prototypes, and design history. Because industrial designers must perform analysis and incorporate manufacturing concepts into their work, classes in communications, mathematics, psychology, physical science, and engineering are also helpful.

Portfolios

A professional portfolio is an extremely important tool for an industrial designer. It is a standard way for designers to demonstrate knowledge, skills, and style when applying for jobs and bidding on contracts. Most schools assist students with putting their portfolios together, helping them select the best examples of their design work from class assignments, internships, and independent projects. Traditionally, portfolios are a collection of sketches, photos, and prototypes or samples. Today, it is also important for industrial designers to have a digital portfolio, which may be presented on a website.

Internships

An internship at a design firm or manufacturing company provides a great opportunity to apply classroom training to a real working environment. Interns typically work as a member of a design team, conducting research, assisting with design creation, and developing prototypes. Most internships are open during the summer, though there are some part-time programs that run during the school year. To find interning opportunities, visit your college career center or check in with your counselor. Be sure to look for internships that closely relate to the field you intend to pursue. For example, if you are aiming for a career in motor vehicle design, an internship at General Motors would be ideal. Aspiring package designers, on the other hand, should look for opportunities with manufacturers of retail products. Companies like Hershey's, Hasbro, or Johnson & Johnson would be good choices.

Graduate School

A growing number of industrial designers pursue master's degrees. Some are traditional art school graduates who have earned BFAs, and want to earn a master's degree in industrial design. A smaller number are interested in the technical side of industrial design. For them, the goal is a graduate degree in engineering. There is a trend toward developing business skills. In general, employers like to see industrial designers with knowledge of marketing, quality control, accounting, project management, and especially strategic planning. Earning a master's degree in business administration (MBA) is the best way to learn how to strategically design according to a company's business plan. It also helps a designer qualify for management positions.

On-the-Job Training

It is customary for beginning industrial designers to receive training from their employers. Training programs vary. They may be formal or informal, and last from three months to a few years. Smaller firms are more likely to provide informal training by simply placing the trainee under the supervision of an experienced designer for a few months. Large companies usually have design departments with organized training systems in place. It can take several years for a new designer to learn the various aspects of the job and advance to become a senior designer.

EARNINGS

THE MEDIAN ANNUAL INCOME FOR industrial designers nationwide is about $70,000. Actual earnings may vary widely depending on experience, location, and industry.

Industrial designers at the low end of the scale earn an average of $40,000. These are usually beginners with little or no experience. Those near the top earn more than $100,000. Junior designers, typically those who have been on the job for two or three years, earn around $50,000. Senior designers with solid résumés and 10 years of experience earn more than $85,000 on average, and in some parts of the country it can be more than $90,000. Veteran industrial designers pushing 20 years in the field earn an average salary of $125,000.

After experience, geography is the biggest factor affecting pay. There are about 1,600 industrial design organizations in the US, paying out approximately $1.5 billion in annual payroll. The highest paying employers are located in the state of Michigan, where the average salary is $76,000. Other states where the salary is above average include Massachusetts, New York, South Carolina, Nevada, California, New Hampshire, North Carolina, and Connecticut.

The top industries, paying the highest median annual wages for industrial designers are as follows:

Engineering and architectural services
$70,000

Manufacturing
$63,000

Specialized design services
$62,000

Wholesale trade (firms selling to governments and institutions rather than consumers)
$58,000

Most industrial designers work full time and are therefore eligible to receive benefits. Typical benefits packages include paid vacation time, sick leave, and health insurance with dental coverage. Many also enjoy additional perks such as year-end bonuses, matching 401K plans, and profit sharing. On average, these additional benefits can boost a designer's total compensation about $30,000 a year.

Self-employed designers, owners of consulting firms, and those who work for service firms that hire them out to other companies, have the potential to earn the most. However, for these independent professionals, incomes fluctuate as clients come and go. In addition, they must provide their own insurance.

OPPORTUNITIES

EMPLOYMENT IN THIS FIELD IS projected to grow slowly in the coming years. It is expected that declining employment in the manufacturing industry is the main cause of the lower numbers. However, consumer demand for innovative products and new product styles will sustain the need for industrial designers in general. According to the NEA (National Endowment for the Arts), too much weight is attributed to the manufacturing sector. The NEA asserts that industrial design is at an all-time high, with US design patents reaching a 25-year peak. Over the next few years, the NEA expects that employment for industrial designers in the professional services sector (engineering and specialized design firms) will increase almost 30 percent. This is a fairly limited sector, but it should still be able to offset some of the decline in manufacturing.

Many of the most highly qualified and successful designers get their work from independent specialized consulting firms. Manufacturers are increasingly contracting with these outside firms to do their design work in an effort to cut costs without sacrificing quality. This is where the best job opportunities are – but only for the top designers. These service firms look for the best and brightest talent, preferably with strong engineering backgrounds and excellent CAD design skills.

Choosing a hot specialty can make all the difference when setting out on this career path. The most active product categories include furnishings; recording and communications; tools and hardware; packaging; food service equipment; transportation; environmental heating and cooling; and games, toys, and sports. It is anticipated that industrial designers specializing in consumer electronics, medical equipment, and transportation will experience very high demand. Of these, the strongest job outlook for the foreseeable future is in medical equipment. Employment of industrial designers who design precision instruments for all kinds of laboratories is also likely to continue to grow. Both areas require a high degree of technical ability and design sophistication.

The continuing emphasis on product safety as well as quality is driving overall demand for new product designers. The best prospects are those who can design new products that are easy to use, safe to operate, and inexpensive to manufacture.

Green design has been a growing trend for more than a decade. The increasing focus on the use of sustainable resources is likely to improve prospects for designers with the knowledge to design earth friendly products. This is not a small niche today, and it is growing. Tesla's Powerwall and PowerPack (nontoxic battery storage systems) are game changers that will make it possible for other green technologies such as electric cars to move squarely into the mainstream. Other examples of green products range from designs that can reduce littering to wind-powered farm equipment.

Advancement Opportunities

New industrial designers can expect to start moving up to higher-level positions after two or three years of on-the-job training and experience. Generally, the larger the employer, the more opportunities there will be for advancement. The first level up is usually to team leader. From there, an experienced designer can become chief designer or design department head, or move into other supervisory positions. Those who work in small companies have fewer opportunities for advancement with their employers. They are most likely to move on by going out on their own and opening their own design firms. Some designers become teachers in design schools or in colleges and universities. Teaching is usually a part-time activity, supplemented by private consulting or operating small design studios on the side.

GETTING STARTED

PREPARING FOR A JOB SEARCH STARTS the moment you finish your first school project. Collect and carefully store your best-completed projects in folders, which will later be transformed into a portfolio. A professional portfolio is an industrial designer's most important job search tool. It serves as an illustrated résumé, documenting your talent, style, and skills. As graduation time approaches, go through your collection and cull your very best work. You can include anything that is relevant: hand drawings, photos, computer images, and physical samples. Keep in mind that portfolios are adjustable. You can (and should) adapt your portfolio contents to appeal to each prospective employer.

Add a written résumé to your portfolio and start looking for interviews. Attend as many as you can. The more interviews you do, the better you will get at presenting yourself and answering questions. How do you get interviews? There are several good ways to get in front of a prospective employer.

Your college career center or job placement office should be your starting point. Most employers prefer to hire college graduates so it is only natural that they look on campus for candidates. The career center keeps a list of job openings and also posts notices of upcoming job fairs and recruiters that will be visiting.

Conduct regular job searches online. You can find industrial design jobs on general job boards, but there are numerous websites dedicated to the field of design. Look for job boards that focus only on jobs for industrial designers.

Keep in touch with your industry. Join professional organizations like the Industrial Designers Society of America. Attend meetings and seminars whenever possible and learn of opportunities. Read professional journals and trade magazines. This is a great way to learn about new or expanding companies that may be hiring more design staff.

Cold call. Many (if not most) great job opportunities in the design field are passed by word of mouth long before they are advertised. Applying directly to companies that hire industrial designers is the best way to land the kind of job you really want. There are several directories that can make this search easier. *The Design Firm Directory*, for example, publishes a graphic and industrial design edition that is published yearly. There are also many trade show directories that list companies within various industries such as toys, housewares, medical products, etc.

Keep your list of contacts up to date so that you can keep in-touch with prospective employers. Make a point to regularly add to your list by attending talks and activities related to industrial design. Look for opportunities to interact with other designers who already have jobs to get their inside scoop on possible openings. Employers of industrial designers get inundated with résumés, especially around graduation time. Keep your résumé from getting lost in the pile by staying in touch. By calling up every few months, you will be remembered when a job opportunity does arise. Do not be concerned with being a nuisance. Companies like enthusiastic employees who demonstrate a strong desire to get down to work.

Take any job you can find – as long as it is related in some way to industrial design. The product development process is long and complex, and there are many kinds of jobs involved. For your experience to carry any weight though, you must make sure to clearly describe in your résumé how every job you did contributed to your abilities in this career.

ASSOCIATIONS

■ **Industrial Designers Society of America**
www.idsa.org

■ **International Council of Societies of Industrial Design**
http://www.icsid.org

■ **National Association of Schools of Art and Design**
http://nasad.arts-accredit.org

■ **Association of Women Industrial Designers**
http://www.awidweb.com

■ **International Furnishings and Design Association**
http://www.ifda.com

■ **User Experience Professionals Association**
http://uxpa.org

PERIODICALS

■ **Design News**
www.designnews.com

■ **Core77**
http://www.core77.com

■ **I.D. Online**
http://idonline.com

■ **Design Journal**
https://www.designjournalmag.com/the_magazine.htm

■ **Innovation Journal**
www.innovationjournal.org

WEBSITES

■ **The Design Firm Directory**
www.designfirms.org/directory

■ **Design-Engine.com**
http://www.design-engine.com

■ **Coroflot**
http://www.coroflot.com

■ **Dribble**
https://dribbble.com

■ **Rhode Island School of Design**
http://www.risd.edu/industrial.cfm